how2become

How to Become

an

Australian Firefighter

www.How2Become.com

As part of this product you have also received FREE access to online tests that will help you to become a Firefighter

To gain access, simply go to:

www.MyPsychometricTests.co.uk

Get more products for passing any test at:

www.How2Become.com

Orders: Please contact How2Become Ltd, Suite 14, 50 Churchill Square Business Centre, Kings Hill, Kent ME19 4YU.

You can order through Amazon.co.uk under ISBN: 9781912370085, via the website www.How2Become.com or through Ingram.

ISBN: 9781912370085

First published in 2017 by How2Become Ltd.

Typeset by Katie Noakes for How2Become Ltd.

Disclaimer

Every effort has been made to ensure that the information contained within this guide is accurate at the time of publication. How2Become Ltd is not responsible for anyone failing any part of any selection process as a result of the information contained within this guide. How2Become Ltd and their authors cannot accept any responsibility for any errors or omissions within this guide, however caused. No responsibility for loss or damage occasioned by any person acting, or refraining from action, as a result of the material in this publication can be accepted by How2Become Ltd.

The information within this guide does not represent the views of any third party service or organisation.

Table of Contents

Introduction

The history of Australia is fascinating, and the development of the emergency and firefighting services in the country are just as unique as any other part of the nation's development. What began as many small volunteer groups has developed into a very organized system for fighting bush, home, and structure fires as well as many organizations devoted to education and rescue services.

Today, there are hundreds of stations across Australia's six states and two territories. Firefighters are perceived as working in three specific "theatres". There are urban, rural and government land firefighters under the authority of a specific entity or group. This means that Australia's population of roughly 22 million people and almost three million square miles of terrain are well served by many salaried and/or volunteer firefighters in these regions.

Anyone seeking a rewarding and exciting career in firefighting will want to explore the many possibilities that Australia presents. The income potential is excellent, the benefits many, and the work always interesting and important. So, how do you get a position in one of the fire companies? What is required from a candidate? Do you need a special educational background?

This guide will present you with the details required to get started in a firefighting career in Australia. In the following pages you will learn a bit about firefighting and how it developed in Australia. You will also learn how emergency and firefighting teams have become essential parts of each community and discover just how to apply for the job of your dreams.

Chapter One

Firefighting in Australia

Encompassing a total area of more than two million square miles, Australia is enormous. It has a population of approximately 24 million people. However, the population is not evenly distributed, with the majority of its inhabitants located along the coast. Australia is considered to be one of the most developed nations in terms of infrastructure and its economy. In fact, it has the world's fifth highest per capita income and is one of the largest economies too.

Naturally, this means that Australia's government organizations are well established, which is impressive since it has only been an independent nation since 1901. It has evolved rapidly, and continues to do so. This can be seen in all parts of society, including how communities and cities handle emergencies.

This chapter will take a quick look at the ways the major areas of Australia have learned to deal with fire and rescue emergencies, and where employment opportunities are readily available.

A Brief Explanation

Australia's different firefighting organizations have evolved rapidly in Australia. Over 100 years ago, the six colonies of Australia federated into the Commonwealth. Today, there are six states and two mainland territories:

- New South Wales;

- Queensland;

- South Australia;

- Tasmania;

- Victoria;

- Western Australia;

- Northern Territory;

- Australian Capital Territory.

Over the years, each of these regions has found it necessary to develop firefighting organizations according to its own particular requirements. Though some of the groups would initially take their cues from outside influences, the development of the actual brigades and companies resulted from the needs of the communities in which they operated.

This is why you will find that some areas rely entirely on volunteer services, yet others have fully salaried teams of firefighters on call on a 24-hour basis. Some crews are experts at fighting bush fires, and others know the best ways to combat urban structure fires.

Moreover, large swaths of the Australian continent are sparsely populated, yet still pose a serious threat for major bushfire incidents. Because it is too costly for all towns, villages, shires or hamlets to have a full time/permanent/paid fire service, the government subsidizes firefighting operations.

For instance, the government will purchase equipment and pay for training, but it is in the hands of the volunteers to use the gear and their skills accordingly. Thus, you can get full training as a volunteer or as a permanent firefighter.

This means that there are many options for someone interested in becoming a firefighter in Australia.

Before going into details about applying for a position, it is worth taking a look at the places where work is to be found in the different regions. Take note that each group has its own unique approach to hiring any salaried workers, and the basic or essential steps for each organization will be identified in a later chapter.

For the time being, different entities or organizations within each regional firefighting group will be highlighted, and it will also be noted when any group only offers volunteer opportunities.

New South Wales

NSW Rural Fire Service

This is Australia's very first official bush fire brigade, banded together in Berrigan around 1896. Firefighters in this brigade worked in organized patrols to keep an eye on any bush fire conditions. There had been deadly, uncontrollable bush fires in Northern Victoria and southwestern New South Wales in the early 1890s, and this group established itself and trained to combat the issue.

The volunteer brigade system continued until the 1930s when the Bush Fires Act gave local councils the authority to establish brigades and appoint officers within them. This evolved into even more distinct groups when the Bush Fires Act of 1949 was passed. By 1970, the Bush Fire Committee had been created out of twenty different groups that include local government, insurance companies, farming groups, and more. Its goal was to advise local government on anything related to bush fires.

By the 1990s, the Bush Fire Committee had developed even farther with the Committee first becoming the Bush Fire Council, and later the Bush Fire Service. It would coordinate firefighting activities throughout the region, but the brigades were still directly controlled by local councils.

In 1997 the NSW Rural Fire Service was created out of the Bush Fire Service and is an eight region, state-controlled entity. The district offices manage the local brigades and develop the fire prevention strategies, and currently there are more than one hundred offices.

Fire and Rescue NSW (FRNSW)

With almost seven thousand firefighters, six thousand volunteers and 414 paid support staff, the Fire and Rescue NSW is the seventh largest urban fire service in the modern world. It works directly with the Rural Services, but originated as the Metropolitan Fire Brigade from Sydney founded in 1884.

By 1909, it become its own organization: the New South Wales Fire Brigades, and in 2011 it became the Fire and Rescue NSW to better describe what the group provides in terms of community services.

Within this organization you can find a range of employment opportunities due to the presence of permanent, retained and mixed stations – all of which are outlined below:

- Permanent Stations - Manned with full time firefighters who work on a rotating shift of four different "platoons".

- Retained Stations - Manned by part-time firefighters who have received the same training as full-time firefighters but who usually have a regular job apart from firefighting. These are stations in which firefighters are "on call" and respond from work or home 24-hours a day.

- Mixed Stations - These can vary widely and will have a blend of permanent and retained workers. They may have two groups of equipment that are used by the different staff members as well, meaning a permanent "first response" team and a secondary support group with additional vehicles available as needed.

NOTE: This is a format used by almost all of the fire services in Australia. If a brigade is made up of retained firefighters, this means that the staff are fully trained and working "on call", etc.

This organization also has Community Fire Unit Volunteers who are residents with special training in bushfire prevention and tactics. They are trained especially to help reduce the effects of bushfires within their own communities. These volunteers are trained at their local stations and focus on bushfire prevention, preparedness and education.

Remember that all of the Fire and Rescue NSW groups are meant to deliver more than just firefighting services. They are also rescue and "hazmat" or hazardous materials providers. FRNSW is under the leadership of the Government of New South Wales and receive the majority of their funding from them.

Queensland

Queensland Fire and Rescue Service

Among the oldest of the firefighting organizations in the nation, the Queensland Fire and Rescue service was founded in 1860 after a fire destroyed a major workshop in the city of Brisbane. By 1889, the city had a paid brigade employed by the city, though there were already many well established laws and guidelines about fire.

Until 1990, the organization existed as a group of 81 local Fire Boards, and was consolidated into three different councils. By 1997 it had become the Queensland Fire and Rescue Authority, and in 2001 was renamed Queensland Fire and Rescue Service.

It has 2100 full-time (permanent) offices and another 2100 part-time (retained) personnel. The Rural Fire Service handles all of the rural, semi-rural and "fringe" areas.

Rural Fire Service Queensland

Although the Queensland Fire and Rescue covers the urban areas of the state, the Rural Fire Service actually provides firefighting services to around 93% of Queensland proper. This is why there are usually more than 34,000 volunteers and more than 2,000 fire wardens necessary to ensure complete coverage.

This organization does have around 750 paid "service employees" which includes administrative and maintenance staff among them.

South Australia

Metropolitan Fire Service (SAMFS)

Established in 1867, the SAMFS is based in the capital city of the region - Adelaide. It requires more than 1000 staff members in the 36 stations to remain functional. It has permanent and retained firefighters, and even uses one marine vessel.

It is a government funded service (since 1981) that focuses on reducing the impact of fires and emergencies in South Australia, though most of the state is served by the Country Fire Service.

Country Fire Service

This is a volunteer service that covers more than 400 communities in South Australia. It attends to more than 8,000 emergency incidents (including bushfires, automobile accidents and hazmat issues) each year. It relies on more than 13,000 volunteers and has only 110 paid staff members.

Tasmania

Tasmania Fire Service

Until 1979, firefighting in the small island State of Tasmania was done by 22 Urban Fire Brigades, a Rural Fire Board and the State Fire Authority. After that time it was consolidated into the Tasmania Fire Service.

Today it has 230 brigades, more than 4,800 volunteer firefighters and a paid staff of roughly 250 personnel. This group is responsible for managing and controlling fires in the state, but also in some of the smaller surrounding islands.

In addition to firefighting services, the teams are also trained in hazmat operations, rescue services and urban search. To accomplish this list of tasks, the TFS cooperates with other emergency service providers that include the Police and Ambulance corps, as well as the Parks and Wildlife Services.

Victoria

Country Fire Authority

The State of Victoria is home to one of the country's largest cities - Melbourne. The CFA or Country Fire Authority, however, is not focused on this region. Instead, it provides emergency and firefighting services to the rural and suburban regions. Currently it has eight regional headquarters, each of which has 20 districts. This means that there are roughly 160 districts operating throughout Victoria, in addition to the MFB or Metropolitan Fire Brigade which focuses on Melbourne.

The CFA is the result of years of struggling to determine the best ways to fight bushfires. After a series of fires in 1944, several commissions and groups were merged into the CFA. It is now the largest volunteer emergency organization in the entire world.

There are opportunities for **both** volunteering and working as a career firefighter in this organization. Currently there are more than 1,400 paid staff members with more than 500 of them being paid firefighters.

Metropolitan Fire Brigade (MFB)

One of Melbourne's oldest organizations, the MFB began operating in the city in 1845. It was formally established in 1891 and already had 56 individual brigades. Over the years the group worked to develop enhanced water delivery systems and improved monitoring services. By the 1950s it had grown large and disorganized, which prompted a conversion to a "platoon" system that used a 10/14 shift structure.

By then, the firefighters were also providing rescue, hazmat and other emergency services throughout the entire metropolitan Melbourne region.

Today, there are 54 stations spread equally around four zones in the city. The Central, Northern, Southern and Western zones rely on more than 1,700 firefighters working 24-hours a day. There are also corporate and part-time positions in the MFB.

Department of Environment, Land, Water & Planning

Formerly known as the Department of Sustainability & Environment (DSE), the DEWLP was founded in 2015 from a merge of the DSE and the Department of Environment and Primary Industries (DEPI). It recruits seasonal "project firefighters." These are employed full-time for a fixed term each year at more than 80 sites in Victoria.

Project firefighters are tasked with combatting bushfires, as well as preventative operations such as planned burning.

Western Australia

Inhabited by roughly 2.3 million people, the Western Australia region is the second largest sub-national entity in the world. Most of the inhabitants of this enormous geographic region live in the southwestern corner of the state - surrounding the city of Perth.

This does not mean that the million-plus square miles of remaining terrain can be ignored. Bushfires could have severe consequences for communities, and for most of the 19th Century and early 20th Century, firefighting in this region was insufficient.

In the 1960s, the Australian Civil Defence Service was established to combat potentially disastrous issues such as bushfires and accidents. That group became the State Emergency Service (SES). It employs more than 1,000 career firefighters and relies on over 32,000 volunteers. They will operate under a range of different groups.

For example, the Fire and Emergency Services Authority of Western Australia (FESA) was established in 1999 (with government support) to administrate most of the emergency services, including:

- Fire and Rescue Service (both paid and volunteer);

- Bush Fire Service (both paid and volunteer);

- State Emergency Service.

These are divided between country career fire stations and metropolitan career fire stations. (In this case, the term "career" is going to mean permanent or full-time firefighting work).

Northern Territory

Northern Territory Fire and Rescue Service

The expansive Northern Territory contains almost 1.5 million square miles and is inhabited by less than 230,000 people. Roughly half live in the city of Darwin and the rest in smaller towns and villages. Fire and rescue services are provided by the Northern Territory Fire and Rescue Service (NTFRS). This group works alongside the volunteer bushfire brigades from Bushfires NT.

The NTRFS is relatively young since it only originates in the 1940s. Until then, the inhabitants of the entire territory would form bucket brigades in the event of a fire. Shortly before the Second World War reached Australian shores, a major fire in the city of Darwin led to the establishment of a more organized fire protection system. When the actual threat of war was present the city decided that a fire service was essential.

This newfound fire service fared well, but by 1958 changes were implemented and a more modern service emerged. With increasing population, more services and stations were needed. Today, there are 26 stations (around 16 are staffed with volunteers), with 4 manned on a permanent basis. The rest of the stations are a blend of permanent and volunteer groups (with some retained/auxiliary fighters on hand too).

In total, there are around 180 permanent firefighters, 54 retained fighters and more than 250 volunteers.

Bushfires NT

This is a Government sponsored entity tasked with finding ways of reducing the total area burnt by wild/bushfires in the territory. In order to do this, Bushfires NT, develop management and response plans, educate the public, create mutual aid plans, and much more.

Anyone interested in learning about rural and bushfire control can enlist as a volunteer in one of their brigades.

Australian Capital Territory

Australian Capital Territory Fire Brigade

What started in 1913 as the Canberra Fire Brigade has developed into the Australian Capital Territory Fire Brigade. It had a relatively slow growth for many years, increasing only to five paid firefighters and three vehicles by 1931. Within the next thirty years, however, the city's population increased so dramatically that a second station was needed.

Today, there are 10 stations which serve as a part of the Australian Capital Territory Emergency Services Agency. The firefighters cover a tremendous list of services that range from structure fires to hazmat, rescue, and even fire investigation.

ACT Rural Fire Service

The ACTRFS is the rural extension of the fire service under the ACT Emergency Services Agency. This is purely a volunteer group that holds itself responsible for protecting lives, property and the environment from bush and grass fires in the Australian Capital Territory.

There are some paid or salaried ways of working for this organization, and it also encapsulates the Territory and Municipal Service employees (TAMS) as well. This means you might volunteer with one of the brigades or find a way into a role in the TAMS.

The Role of the Firefighter

Although we have reviewed the different ways that you might work as a firefighter in Australia, we haven't defined the role of the firefighter in general terms. This is intentional, since in order to discover which role you hope to play in firefighting, you must first have an understanding of how firefighting is done in the different regions and territories.

For example, remember that fire services are viewed in three ways:

- Urban and salaried;

- Rural and volunteer (with some salaried options);

- Government land with both salaried and volunteer opportunities.

Additionally, it is fair to say that each group or organization has its own mission, culture, history, and goals. Thus, potential firefighters have to consider these things as they begin to select the ideal working environment for their needs.

Take the time to consider if you want to work in a rural area, urban landscape, in a full or part time capacity, in a place that looks only at your skill set or one that accommodates for your community mindedness, etc.

Remember too that most firefighters are not going to "just" fight fires. Most of the groups identified above are part of bigger groups and communities. You might be working to educate citizens and children about fire safety, or perhaps help committees to determine the best ways to reduce risk or to prevent fires in the first place. Firefighters are also going to be called upon to do things such as:

- Inspect the aftermath of bush or structure fires;

- Inspect buildings for safety and giving official approval;

- Work in rescue operations of all kinds including auto accidents, marine incidents or even rural/outback issues;

- There may even be a call for disaster response in certain firefighting jobs as well.

This all implies that a lot of skills and experience are required. Fortunately, all firefighting organizations have an educational period in which their fighters are trained, but you still need that "extra" something.

What is that extra something? Most successful firefighters tell you that it can include a range of things such as:

- Enjoying being part of a team;

- Having a strong commitment to the community;

- Being agreeable to constant and ongoing training;

- Being able to work directly with the public in many different ways such as education, training, and preparedness;

- Being focused on preservation of wildlife and natural habitats;

- Being able to operate independently but to also quickly follow orders from a higher authority figure;

- To remain calm in any sort of crisis or emergency situation.

Does this sound like you? If so, you are probably a good candidate for employment as a firefighter of some kind in Australia. Let's move into some of the more specific requirements...

What it Takes to be a Firefighter in Australia

If you have been reading this guide, you're clearly more than just interested in becoming a firefighter. Having a passion for the role is necessary for becoming a firefighter. Passion is valuable, but there are other requirements that you will need to meet. Some of the requirements are very basic, such as having Australian citizenship or permanent residence status, but a brief list of things to consider includes the following:

- Australian citizenship or permanent residence status (New Zealand citizens may also apply);

- An education to the minimum level acceptable by the organization (for the most part Year 10 is acceptable, but most organizations prefer to see Year 12 levels); OR a completed Trade Certificate or Tertiary Qualification;

- The ability to demonstrate considerable employment experience (usually this means at least two years of full time employment);

- Current first aid certificate, particularly the HLTAID003;

- Australian C Class Licence with no restrictions and confirmation of completion of a 'Heavy Vehicle Knowledge Assessment', OR an Australian MR Class driving licence with no restrictions (some services permit firefighters to obtain their Heavy-Rigid license within the first year of employment);

- Current Medical Consent to participate in physical abilities testing;

- A police character check.

If you know that these are things that will not present you with any challenges, you can move on to the next chapter. If you have some gaps in your requirements, take the time right now to fill them in because these are essential items for almost any paid or salaried firefighting position.

Be Sure to Fill in the Gaps

If you did not complete all 12 years of secondary education, get a certificate. Alternatively, work towards a tertiary certificate. If you are younger and have not yet been employed for two full years, consider if there is anyone in the community who might recommend you because of your reliability. Letters of recommendation are always beneficial, especially if you are short on some of the requirements.

However, there is one thing that is going to be nearly impossible to overcome or "explain away": a history of criminal activity. Few firefighting organizations will accept the application of someone with a long criminal record. If you do happen to have a slightly shaky history, try to find individuals who will give you a written reference as witness to your good behaviour.

Also remember that most firefighting jobs are physically demanding and the organizations have minimal health limits that they will accept. Therefore, it is important that you maintain a fitness regimen if you aren't in the best shape. For example, the "Beep Test" usually has a 9.6 minimum required. Moreover, some medical conditions such as asthma, degenerative arthritis and hearing/vision loss may prevent you from becoming a successful candidate.

Keep in mind that all firefighting groups are equal opportunity employers. Some even make their organisational procedures or policies available online.

This means that discrimination based on age, career status, disability, gender identity, industrial activity, breastfeeding, marital status, parental status, physical features, political belief/activity, personal association, pregnancy, race, religious belief/activity, sex, sexual orientation/lawful sexual activity will never be allowed to prevent someone from being hired.

These are general requirements across all services, but individual states may have specific prerequisites which you must also meet. In the next chapter, you will learn how and where to apply for a role as a firefighter.

Chapter Two

Applying to be a Firefighter

As we get deeper into the actual process of applying to be a firefighter we suggest that you make a list of deadlines and things "to do". For example, if you see that an organization is not currently recruiting, but that they list the next dates when applications will be accepted, note this on your own agenda.

Keeping an agenda of when services are recruiting will allow you to make use of your time in-between application periods. This is best used strengthening your candidacy by keeping fit and participating in local volunteer work, either with a firefighting (or affiliated service) or as part of the wider community.

The time between recruitment periods is extremely valuable since it will let you focus on improving your chances of being successful. Keeping track of when services are recruiting, as well as whether you meet their requirements, will help you maximise your efficiency during application time. In turn, this will make you a more appealing candidate.

Below is a table of information that you may want to use as you begin preparing for your application process:

Application Worksheet

Use a single worksheet for each application you have or will submit. Fill out the details ahead of time and make sure that you have met all of the deadlines, requirements, etc.

Name of Service	
Address	
Application Completed?	
Application Date	
Next Recruitment Period	
Location Preferences	
Referee 1	
Referee 2	
Referee 3	

Essentials:	Yes	No	Deadline to Resolve Issue
Australian citizenship or permanent residence status;			
An education to the minimum level acceptable by the organization (for the most part Year 10 is acceptable, but most organizations prefer to see Year 12 levels); OR a completed Trade Certificate or Tertiary Qualification;			

The ability to demonstrate considerable employment experience (usually this means at least two years of full time employment);			
Current first aid certificate (many groups want to see the Senior First Aid Level 2 certification);			
Australian C Class Licence with no restrictions and confirmation of completion of a 'Heavy Vehicle Knowledge Assessment', OR an Australian MR Class driving licence with no restrictions;			
Current Medical Consent to participate in physical abilities testing;			

A Police Character Check;			
Notes:			

As we mentioned in Chapter One, each agency has its own approach to hiring firefighters. The best thing to do is to pay a visit to their individual websites to discover what it is that the organization seeks in a good candidate.

For example, one group may recruit based only on ability, knowledge and skill, while another group may prefer candidates who are community-minded. For this reason, it is up to you to find which organizations are the best match for your goals and your skills.

How and Where to Apply

There are many different ways that fire companies accept applications. Some use an online format and others require "in person" meetings. Use the URLs below to pay a visit to the "Volunteer" or "Employment" pages for all of the firefighting organizations mentioned in the previous chapter. Also be sure to:

1. Download and print out any blank applications when requested by the site;

2. Read through any brochures or lists of "tips" that a site provides;

3. Print out any step by step manuals when provided as these are often essential to success;

4. Make a calendar of deadlines (as suggested above). Some groups are not currently recruiting but do make the next dates readily available. This is also a good way to find volunteer training schedules and to use them to "fill in the time" as you prepare for your firefighting application process; and

5. Have a pre-made package of the "essentials" along with a CV (if possible) to facilitate the process.

The Package

One way to be a memorable applicant is to have a pre-made package of the documents and details that are essential. For example, have as many medical certificates as you need on hand, and put them in a file or folder to submit with the application. If you have documents relating to your community volunteerism or activities, include copies. If you have letters of reference, make copies for each packet.

A package with all of the necessary materials shows initiative and diligence, and will also make your potential employer's job much easier during the selection process.

It might be useful to think of this package as a kind of "marketing" package that shows you to be a top candidate. Your dream role is likely very competitive, and so you need to sell yourself as the best candidate for the job. You can reinforce the idea with clippings or images that show you will be a great asset to a firefighting company.

We will get into the details of a good CV in the chapter on interviewing, but for now we do suggest that you begin developing a very simple and concise document that details your work history and your professional goals.

The Australian Firefighting Services

Use the following sites to get all of the information you need to make a proper application for employment as a firefighter or volunteer:

- Australian Capital Territory Fire and Rescue: http://esa.act. gov.au/actfr/careers/new-recruit-requirements/

- Bushfires NT (volunteers only): https://nt.gov.au/ emergency/cyclones/volunteering-cyclones,-fire-and-disasters/volunteer-with-bushfires-nt

- Country Fire Authority (Victoria) (paid and volunteer options): http://www.cfa.vic.gov.au/volunteer-careers/

- Department of Environment, Land, Water and Planning (seasonal project firefighters): http://www.delwp.vic.gov. au/fire-and-emergencies/firefighting-and-employment

- Fire & Rescue NSW (permanent and retained options): http://www.fire.nsw.gov.au/page.php?id=2

- Fire and Emergency Services Authority of Western Australia (FESA) (paid and volunteer options): http://www. fesa.wa.gov.au/recruitmentandtraining/recruitment/Pages/ firefighterrecruitment.aspx

- Metropolitan Fire Brigade (Melbourne): http://www.mfb.vic. gov.au/Recruitment.html

- Northern Territory Fire and Rescue Service (paid and volunteer options): http://www.pfes.nt.gov.au/Fire-and-Rescue/Careers-in-firefighting.aspx

- NSW Rural Fire Service (volunteers only): http://www.rfs. nsw.gov.au/dsp_content.cfm?cat_id=1004

- Queensland Fire and Rescue Service (paid and volunteer options): http://www.fire.qld.gov.au/employment/

- Queensland Rural Fire Service (volunteers only): http://www.ruralfire.qld.gov.au/Volunteering/

- South Australian Country Fire Service (volunteers only): http://www.cfs.sa.gov.au/site/volunteers_and_careers.jsp

- South Australian Metropolitan Fire Service (permanent and retained positions): http://www.mfs.sa.gov.au/site/join_us.jsp

- Tasmania Fire Service (paid and volunteer options): http://www.fire.tas.gov.au/Show?pageId=colJoinTFS

The next chapter provides a summary list of information about the selection criteria that many organizations use. Be sure to spend the time visiting the sites that are the most relevant to you.

While a service's location might be unsuitable for you due to distance, it's worth reading through each agency's details to see if one is a much more suitable solution than another. You will also find a lot of quotes, details and programs just by perusing the sites.

You can, and should, use everything that you have in your favour as you begin to apply to the firefighting services of your choice. The broader your knowledge of the different services, the more informed your decision will be.

Guidelines for a Good CV

Since the selection process for a firefighter is extremely competitive, it's absolutely vital that you have a strong CV to stand out amongst the other applicants. We will discuss the selection process in greater detail in the next two chapters. Now, here are some tips on what you should try to include in your CV:

Full Name:

Address:

Mobile Number:

Email Address:

Career Objective:

Be sure to use this opportunity to give a very clear and concise explanation of your short and long term goals.

Professional Experience:

All of your professional experience to date has to go here, but don't neglect to mention things like giving seminars, volunteering in firefighting work, etc.

Responsibilities:

If you have done work relating to firefighting or appliances, present it here.

Educational Background:

This includes certificates, diplomas and online courses.

Certifications and/or Community Honours:

This is a great place to mention community work or honours presented to you, such as honours or levels in scouts.

Hobbies and Interests:

Personal interests are important, since they demonstrate that you're a human being, not a machine. However, you can use this opportunity to show your interest in exercise, the environment, or other activities and hobbies which might make you a stronger candidate.

References:

Two professional and one personal are the minimum.

This outline should give you an idea of how to make use of all of your experience and interests to sell yourself as a candidate. By doing this, you will hopefully stand out among the other applicants and therefore be more likely to get the job.

Chapter Three

The Selection Criteria

By now, you probably have a good idea of which organization(s) to which you will apply. If not, this chapter can be used to determine the "best fit". Generally, the first step in finding the most suitable organization comes from understanding what criteria are used by employers when choosing from a pool of applicants.

The Importance of Visiting the Websites

Additionally, you may be surprised to learn that you are "tested" on the material. Applicants might be asked to demonstrate their knowledge of the service they are applying to. Moreover, a clear understanding of a firefighter's responsibilities is absolutely necessary, making online resources undeniably invaluable. Hopefully you will have the opportunity to show your knowledge in the interview.

So, not only will the visits to the websites help you to make the best choices according to your wishes, it will also inform you of exactly what is expected from you during the application process, interviews, and the job if you are successful.

The Order of Events

So, what is the "standard" application process like? Generally, the order of events looks like this:

1. Application and all required materials are submitted

2. A preliminary interview may occur

3. All of the tests required by the state have to be taken, including any:

    a. Written tests

    b. Psychological tests

c. Aptitude tests

d. Character analyses

e. Criminal history checks

f. Medical examinations

g. Literacy and numeracy exams

h. Reasoning ability tests

i. Mechanical ability tests

j. Interviews

Tests vary depending on the state, but the physical tests are universal because of the demands of the job. Generally speaking, all applicants have to subject themselves to four separate assessments after entering an application, and these will always include the aptitude, character analysis, fitness and physical assessments. <u>All of these usually have to be paid for by the applicant</u>, and completed according to the department's standards. Some have no order at all, and others have a very particular set of guidelines.

4. The Essential Interview – This will be covered in a later chapter, but only someone who has made it through the tests and application process gets called for the interview. This is always going to be held before a committee and will always be in the region in which you will get the role. Therefore, be sure that the interview takes place in a city or area in which you are willing and able to work.

5. Mandatory training - Whether a volunteer or a paid firefighter, you will be required to enter a program of intensive training. Some are 12 to 16 weeks in duration, some are longer. This training is both physical and mental. You will also begin to learn how to use all of the equipment,

including:

    a.   Fire trucks and how to drive them

    b.   Ladders and how to utilize them

    c.   Breathing apparatus and safety

    d.   The mechanics of fire and firefighting

6.   Enter into a probationary period - If you get through the application and training, you will be invited into a six month (or longer) probationary period that determines if you will be invited to become a permanent firefighter.

Your Chances

Most firefighters are reviewed, assessed and interviewed by a committee. Generally, the committee must adhere to:

- EEO - Equal Employment Opportunity standards that indicate that they cannot refuse to hire, or to terminate employment, based on a long list of factors that can include everything from race and gender to sexual orientation, and more.

Moreover, the committee will be doing the following:

1.   Determining if the applicant has demonstrated an ability to work effectively as part of a team. You may not realize that the questions you answer and the tasks you perform are showing your true nature, but they are. If you are someone who tends to need the spotlight and refuses to work as a team player, you may not get through the interview assessment process because the tests will reveal this about you. If you struggle to co-operate with others in a team, then you might not be suited to the role of a firefighter.

How can I achieve this?

This can be tricky to prepare for since the committee is judging your character, particularly how you work with others. If you are worried that your application won't display this, consider devoting some of your spare time to volunteering or other activities which require teamwork, such as a sport. Building yourself up to be someone who can work as part of a team is vital for working as a firefighter, and adopting the mind-set of a team player will make you a stronger candidate in the application process.

2. Identifying your ability and willingness to learn new skills and solve problems. Before you enter into a training program, you will be tested on these issues. All firefighters have to constantly learn new methods, technology and details about their job. If you are not someone who wants to constantly learn about firefighting and the gear or techniques, this may prevent you from receiving a job offer.

How can I achieve this?

You need to be able to show your knowledge, your ability to learn and retain new information, as well as a passion for continuous learning. In chapters five and six, we will provide you with materials to help prepare for the tests.

In addition to this, use online revision materials to learn about the role, ensuring that you remain up to date with current technology and methods. Moreover, take the time to read sample questions for the tests. These will make you acquainted with the format of the tests and the kind of questions you may be asked, and might help you remain calm during the assessment.

3. Discovering if you have the ability to communicate with a variety of audiences. Remember that the job of a firefighter is not limited to jumping into a truck and heading to the scene of a fire. You will also be asked to participate in community and educational programs, inspections, public gatherings

and more. If you are unable to clearly communicate and remain calm in front of a range of groups and individuals, it is going to present a hurdle to your career.

How can I achieve this?

While public speaking might not be required from you when participating in community programs, you need to be sure that you can perform your duties while surrounded by people. Local community work will also make you better equipped to deal with community programs.

If crowds make you anxious or nervous, it may be worth spending some free time on volunteering and community work so that you become accustomed to larger groups of people. This experience may be helpful if you get to the interview stage, since remaining calm is important to answering questions successfully.

4. Assessing your mental and physical ability to perform operational activities effectively. Since this is a physically demanding job for many companies in Australia, applicants must have demonstrated the required level of fitness to fulfil the duties of the role. You need to be able to carry your gear, air packs, and also perform your duties. If you are unable to do the physical work, they cannot reasonably offer you a position.

How can I achieve this?

As mentioned previously, you should maintain a fitness regimen if you are not physically fit. If you are in good health, still ensure that you meet the minimum requirements specified on a service's website.

5. Examining your ability to perform all duties and operate equipment in a safe and effective manner. While you need to show enthusiasm, you also need to keep the general public safe at all times. A firefighter who drives recklessly

or puts other fighters at risk through his or her own actions is not likely to be offered a position in a department or brigade.

How can I achieve this?

Think hard about whether you are a sensible person who considers the wellbeing of those around them. Also, pay attention to how you behave every day – if it's apparent that you are reckless or oblivious to endangering yourself and others, you will need to change your behaviour to be more careful. Prioritise safety and care in all aspects of your application in order to demonstrate that you have what it takes to be a firefighter.

6. Evaluating your integrity and ability to interpret and apply legislation, policies and processes. Firefighting is dictated by different laws. If you look at the history of any fire brigade, you will see that many of them floundered around as community organizations, and then only flourished once different legislation gave them the authority to serve their purposes. This means you have to show a clear understanding of the laws governing firefighting, and also the laws around fire inspections. Remember that you may have to serve as an inspector at the scene of a fire or accident, and this means you need to know the rules, laws, etc.

How can I achieve this?

Ensure that you have a good understanding of the protocols and laws governing firefighting. It might not be possible for you to know all of the details, but at least showing some knowledge as well as a desire to learn more will improve your chances of success in the application process.

Results

If your application is accepted and you make it through the testing phase, it is likely that you will be called for a very serious interview. This will be held in the region where there is a vacancy and where you will become the new recruit.

If you are chosen for the job, you can expect to enter into your training immediately, and then to begin your duties in the region where the interview occurred. The basic duties for a permanent or retained firefighter will usually include:

- Working with a team while being supervised by a Station Officer;

- Responding to fires, rescues and other emergencies;

- Performing building inspections and calls to fires;

- Following checking procedures to ensure that equipment is prepared for an emergency;

- Participate in station duties, which may include maintenance and cleaning of vehicles and equipment, as well as the facilities themselves;

- Performing duties to the specification of standard operating procedures, as well as tasks assigned by a senior officer;

- Operating vehicles while complying with Traffic Regulations and agency procedures;

- Locating fire alarms, detectors, suppression and building control systems, as well as confirming the status of fire alarms;

- Identifying hazards at the site of an emergency as well as assessing potential hazards;

- Assisting any casualties during emergencies as well as providing emergency care through basic life support techniques in compliance with agency standard operating procedures;

- Processing information in accordance with agency procedures;

- Undertaking training and courses in order to remain up to date with new methods as well as demonstrate competence.

If you believe that you can handle these duties, then it's time to start applying to become a firefighter. In the next chapter you will be provided with tips for completing your application.

BONUS - As part of this guide, we would like you to try out our online training course now for FREE at:

https://www.how2become.com/courses/firefighter/

Chapter Four

Completing the Application

Tips for Completing the Application Form

You should know that almost all of the applications will ask for:

- Personal Details - Name, address, etc.

- Equal Employment Opportunity data.

- Where and how you learned about the job.

- Your Background - Usually including your educational information (be sure that you include the highest level and anything beyond secondary school including seminars or other educational venues), employment details, etc.

- Employment Location Preferences - remember that most firefighting services are more than a single entity. Most have dozens of stations and sites from which they operate. Be sure that you know where you hope to be and indicate first and second preferences if possible (be ready to explain why you made those choices).

- Acknowledgement of Essentials/Requirements - Most applications will ask if you have all of the "essentials" listed in the second chapter.

- Referees - Most ask for three (usually two that are work related and one that is personal). Be sure to provide accurate contact information.

- Medical information and a Medical Statement.

- Declarations and additional details - Usually you are asked to swear that you are giving truthful information and that a failure to disclose details may lead to your application being dismissed.

Top 10 Tips for a Successful Application

Now that the content of an application has been outlined, it's time to turn to some tips which will result in a stronger, more professional application.

1. Leave no Blanks

Firstly, ensure that you leave no blanks on your application form. Try to give as much relevant information as possible so that your potential employers can judge your application well. Leaving blanks will cause hassle for those reading your application, and might make them sceptical about your ability.

2. Make an Accurate Presentation

Be sure that the application is a good reflection of how you want to be seen in the selection process. Don't send a smudged document full of misspellings and empty spaces. Be sure that you are presenting yourself in an organized and focused manner because that is what any firefighting service will want to see right away. Take the time and attention to present your application seriously to prove that you absolutely want the job. A clumsy application might make you look disinterested.

3. Be Yourself

In other words, be sure that your application and supplementary materials show who you are. Often, applicants get so focused on demonstrating what great candidates or firefighters they will be that they overlook the importance of presenting themselves for who they are. This means you need to be clear about your interests, activities, family life, and general goals.

4. State Goals

Firstly, state that you want to be a firefighter. However, it's entirely human to have greater goals and aspirations. Do you have a specific rank, a set of skills, or any other goals that relate to the work? Be sure to mention this in the application and in the interview, since it will demonstrate a greater understanding of how the role can serve you as a human being.

5. Identify your Education

Be sure to first present your formal education, primarily school and higher education if possible. Also include seminars or online classes which have taken place outside of school or higher education. Your education happens inside and outside of a classroom, and you should point out any of the "learning" you have done up until this point in time.

6. Get Previous Experience

This is a bit of a challenge if you don't have a lot of time before the application must be made. This will be much easier to achieve if you have kept a calendar of application dates and made use of your time in between. If you can somehow work with a community firefighting group, a land preservation group, or even a children's scouting group to get firefighting related experience, it will make your application glow.

7. Learn about the Trade

We already pointed out that you can really benefit yourself by gaining information from the different websites and all of the available materials relating to the application process. You should also learn about firefighting in general. Understand the terminology, the tactics, and the latest in gear and appliances because it shows your interest.

8. Use the Internet

There are all kinds of forums and websites that can provide you with ways of learning about job openings that relate to firefighting in Australia. You can even begin to use sites such as LinkedIn that let you mingle with professionals who can steer you in the right direction.

9. Triple Check Your Spelling and Grammar

If possible, have a friend or relative take a look at your application and proofread it. Spelling and grammar errors may not seem like much of an issue, but a firefighter must always pay attention to detail and a number of mistakes in your application will be noticed by the committee judging it.

10. Be Sure to Emphasize what YOU can offer

Your application is the place to point out what you have to offer. Take time to consider which of your life experiences demonstrate your skills and attitude. For example, it would be worth finding examples of instances where you've handled an emergency, worked successfully as part of a team, or recognised someone else's individual needs and acted upon them. Use these experiences to display the best parts of your personality.

What's Next?

If you have followed our advice and made a strong application, it is quite likely that you will be called in for an essential interview. As previously mentioned, this interview will be in front of the official committee. It may be at the same time that other candidates are interviewing, and it will always be held in the region in which you will work.

In these final chapters, we will provide materials to aid you in your preparation for the assessments and interview. In chapters five and six, we will take a look at the written tests in greater detail, as well as offer sample questions for you to practice with. Chapters seven and eight will focus on the interview, giving you some practice with sample questions, as well as some sample answers for inspiration.

Chapter Five

The Assessment

In the section on Selection Criteria we identified many of the essential things that you will need for the application. That list contained the different tests that may or may not be used. In brief these can include:

Written Tests

Known as the written selection tests, these tend to measure critical thinking, problem solving, logic and deductive reasoning. Usually, this will be the Vocational Selection Test (VST). This test is written (pen and paper) with multiple choice questions.

There are two parts to the VST:

1. The core VST which is comprised of:

- Verbal Reasoning – Written passages will be offered which will test your ability to read and interpret them.

- Quantitative Reasoning – This section will involve math questions measuring your ability to make deductions based on your mathematics skills.

- Abstract Reasoning – Questions in this section will involve patterns, shapes and sequences. You will be assessed on your ability to identify these patterns.

2. VST - Mechanical Reasoning

- This test will require you to demonstrate an understanding of the basic functions of mechanical devices.

Generally speaking, services will require a score of at least 50 (out of 100) in order to be considered for the role.

Psychological Tests

Most firefighting organizations are going to rely on the AIFP profiling test to get an accurate idea of the candidate's capabilities for mentally handling the stress or strain of the job.

This test is designed to measure your problem solving skills, and your attitude to teamwork, amongst other personality traits. The AIFP can be used to determine whether you are trying to portray yourself as 'perfect'. No matter how great a candidate you may be, you are not perfect, and the committee is not expecting perfection. Be honest in answering these questions because attempting to look perfect will negatively impact your test results. The best advice for this test is to be the best version of yourself you can possibly be – you don't need to be perfect.

Aptitude Tests

This is usually a combination of physical and mental work. You are asked to demonstrate mechanical abilities with firefighting appliances and to show that you have the aptitude for using the gear in a physical manner. If you do not pass on the first attempt, you usually have to wait six months before trying again.

Generally you will see the following format:

- Physical Fitness Test (Multi-stage Shuttle Run)

    o   Must achieve a level of 9.6

- Firefighting Aptitude Tests

    o   Part 1: Ladder climb

    o   Part 2: Firefighting Task Course

You can get preparation guides for the Physical Fitness test from the Australian Sports Commission.

Character Analyses

This can be a group and/or individual assessment. You will participate in things as simple as personal introductions, group discussions, presentations and problem solving.

You may also be asked to write or orally communicate a few simple concepts to present to a group. This will be used to show how well you work as part of a team.

Any such tests are not meant to make you uncomfortable, feel foolish, or trigger a strong emotion. What they are meant to do is allow the observers to determine if you have the interpersonal skills necessary to manage the stress of the role.

BONUS - As part of this guide, we would like you to try out our online training course now for FREE at:

https://www.how2become.com/courses/firefighter/

Chapter Six

Sample Test
Questions

In this chapter, you will have the opportunity to familiarise yourself with the content, language and format of the tests. Remember that some of your tests will be taken under timed conditions, so it might be worth timing yourself when attempting these questions. This way you can make sure you have the time to answer every question in the actual exam.

Verbal Reasoning

Exercise 1

Barry and Bill work at their local supermarket in the town of Whiteham. Barry works every day except Wednesdays. The supermarket is run by Barry's brother Elliot who is married to Sarah. Sarah and Elliot have 2 children called Marcus and Michelle who are both 7 years old and they live in the road adjacent to the supermarket. Barry lives in a town called Redford, which is 7 miles from Whiteham. Bill's girlfriend Maria works in a factory in her hometown of Brownhaven. The town of Redford is 4 miles from Whiteham and 6 miles from the seaside town of Tenford. Sarah and Elliot take their children on holiday to Tenford twice a year and Barry usually gives them a lift in his car. Barry's mum lives in Tenford and he tries to visit her once a week at 2pm when he is not working.

1. Which town does Elliot live in?

A. Redford

B. Whiteham

C. Brownhaven

D. Tenford

E. Cannot tell

2. On which day of the week does Barry visit his mother?

A. Cannot tell

B. Monday

C. Tuesday

D. Wednesday

E. Thursday

3. "Bill and Maria live together in Brownhaven."

A. True

B. False

C. Cannot say

Exercise 2

Janet and Steve have been married for 27 years. They have a daughter called Jessica who is 25 years old. They all want to go on holiday together but cannot make up their minds where to go. Janet's first choice would be somewhere hot and sunny abroad. Her second choice would be somewhere in their home country that involves a sporting activity. She does not like hill climbing or walking holidays but her third choice would be a skiing holiday. Steve's first choice would be a walking holiday in the hills somewhere in their home country and his second choice would be a sunny holiday abroad. He does not enjoy skiing. Jessica's first choice would be a skiing holiday and her second choice would be a sunny holiday abroad. Jessica's third choice would be a walking holiday in the hills of their home country.

1. Which holiday are all the family most likely to go on together?
A. Skiing

B. Walking

C. Sunny holiday abroad

D. Sporting activity holiday

E. Cannot tell

2. If Steve and Jessica were to go on holiday together, where would they be most likely to go?

A. Sunny holiday abroad

B. Skiing

C. Cannot tell

D. Sporting activity holiday

E. Walking

3. Which holiday are Janet and Steve most likely to go on together?
A. Cannot tell

B. Walking

C. Sporting activity holiday

D. Skiing

E. Sunny holiday abroad

Exercise 3

FLIGHT A outbound leaves at 8am and arrives at 1pm. The cost of the flight is $69 but this does not include a meal or refreshments. The return flight departs at 3am and arrives at its destination at 8am.

FLIGHT B outbound leaves at 3pm and arrives at 8pm. The cost of the flight is $97 and this includes a meal and refreshments. The return flight departs at 1pm and arrives at its destination at 5pm.

FLIGHT C outbound leaves at 4pm and arrives at 10pm. The cost of the flight is $70 but this does not include a meal or refreshments. The return flight departs at 10am and arrives at its destination at 4pm.

FLIGHT D outbound leaves at midnight and arrives at 3am. The cost of the flight is $105, which does include a meal and refreshments. The return flight departs at 3pm and arrives at 6pm.

FLIGHT E outbound leaves at 5am and arrives at 12 noon. The cost of the flight is $39, which includes a meal and refreshments. The return flight departs at 5pm and arrives at its destination at midnight.

1. You want a flight where the outbound flight arrives before 2pm on the day of departure. You don't want to pay any more than $50. Which flight do you choose?

A. Flight A

B. Flight B

C. Flight C

D. Flight D

E. Flight E

2. You don't want to pay any more than $100 for the flight. You want a meal and the outbound departure time must be in the afternoon. Which flight do you choose?

A. Flight A

B. Flight B

C. Flight C

D. Flight D

E. Flight E

3. You want a return flight that departs in the afternoon between 12 noon and 6pm. The cost of the outbound flight must be below $100 and you do want a meal. The return flight must arrive at your destination before 6pm. Which flight do you choose?

A. Flight A

B. Flight B

C. Flight C

D. Flight D

E. Flight E

Quantitative Reasoning

Question 1

Which of the following is the same as 25/1000?

A. 0.25

B. 0.025

C. 0.0025

D. 40

E. 25000

Question 2

Is 33 divisible by 3?

A. Yes

B. No

Question 3

What is 49% of 1100?

A. 535

B. 536

C. 537

D. 538

E. 539

```

```

Question 4

One side of a rectangle is 12cm. If the area of the rectangle is 84cm², what is the length of the shorter side?

A. 5cm

B. 6cm

C. 7cm

D. 8cm

E. 9cm

```

```

Question 5

A rectangle has an area of 8cm². The length of one side is 2cm. What is the perimeter?

A. 4cm

B. 6cm

C. 8cm

D. 10cm

E. None of these Answers

Abstract Reasoning

Exercise 1

Complete the pair using the first pair to help you.

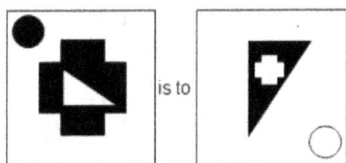

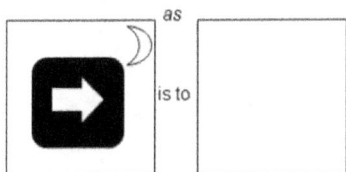

 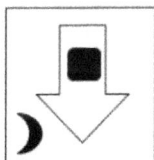  

| A | B | C | D |

Exercise 2

Which figure completes the sequence pattern?

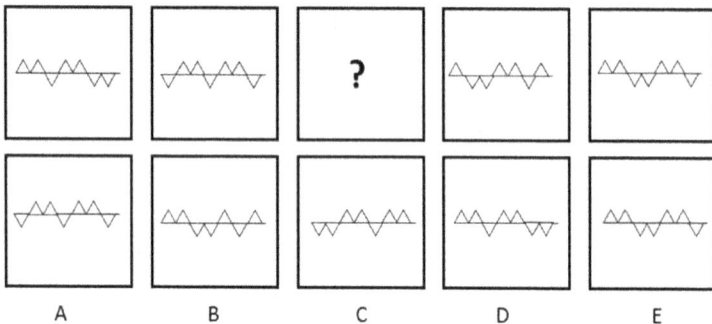

| A | B | C | D | E |

Exercise 3

Which figure completes the sequence pattern?

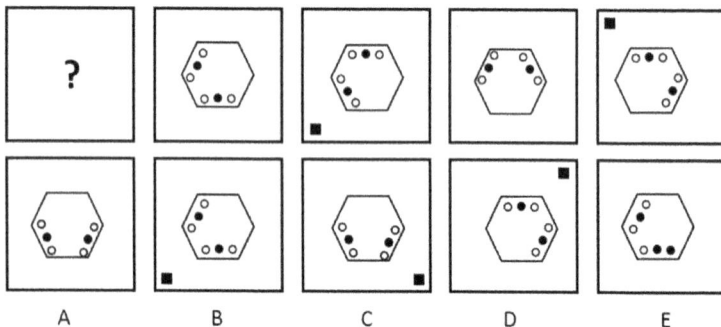

| A | B | C | D | E |

Exercise 4

Using the vertical mirror line, which answer shape is a reflection of the example shape?

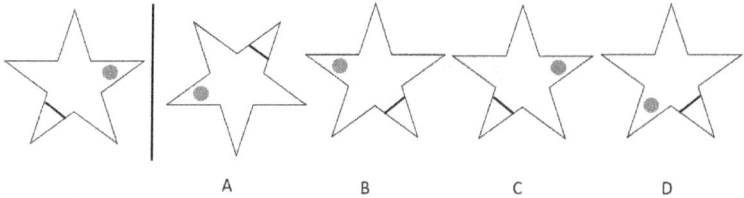

A   B   C   D

Exercise 5

Which answer shape is a rotation of the example shape?

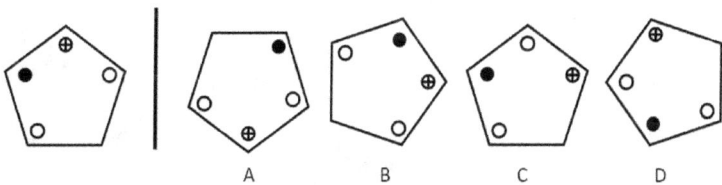

A   B   C   D

Psychological Test

1. I tend to embrace change at work.

Strongly agree

Agree

Neither agree nor disagree

Disagree

Strongly disagree

2. I would take just as much care and attention when mopping the fire station floor as I would servicing my breathing apparatus set.

Strongly agree

Agree

Neither agree nor disagree

Disagree

Strongly disagree

3. Different cultures of people in the community are not beneficial to society.

Strongly agree

Agree

Neither agree nor disagree

Disagree

Strongly disagree

4. Being organized is not important to me.

Strongly agree

Agree

Neither agree nor disagree

Disagree

Strongly disagree

5. There should be more men in the Fire Service than women.

Strongly agree

Agree

Neither agree nor disagree

Disagree

Strongly disagree

Answers to Sample Questions

<u>Verbal Reasoning</u>

Exercise 1:

1. B

2. D

3. C

Exercise 2:

1. C

2. A

3. E

Exercise 3:

1. E

2. B

3. B

<u>Quantitative Reasoning</u>

Question 1: B

Question 2: A

Question 3: E

Question 4: C

Question 5: E

<u>**Abstract Reasoning**</u>

Exercise 1: C

Exercise 2: C

Exercise 3: C

Exercise 4: B

Exercise 5: B

If you pass all of the tests, you are going to have to tackle the interview. By that point, you should already have a clear idea of how to present yourself. In the next chapter, we will focus on how to prepare for the interview, as well as what to expect during it.

BONUS - As part of this guide, we would like you to try out our online training course now for FREE at:

https://www.how2become.com/courses/firefighter/

Chapter Seven

The Interview

Preparing for the Interview

This final stage of the selection process is often considered to be the most nerve-wracking. Moreover, to make it this far and then be unsuccessful due to the interview can be incredibly frustrating. This is why the entire chapter will be devoted to preparing you for the interview.

In this chapter, you will learn what kind of general questions to expect from the committee during your interview. You will be supplied with sample questions as well as some sample answers to give a general idea of how to answer during the interview.

In addition to this, we will cover some general advice on how to physically prepare for the interview in order to be as successful as possible.

The first thing to note is that you are certainly due to face a committee. Try not to feel too daunted by this – you have done just as well as the others in the application process so far, and so there's no reason to be intimidated. If you've followed the advice in this guide, you're going to be fairly knowledgeable about the application process and the interview.

It's completely natural to find yourself nervous about the interview. In fact, it shows that you truly want the role. If you didn't care, you wouldn't be nervous. However, simple breathing and mental relaxation exercises can help to keep your nerves at a manageable level. Preparation will also make you slightly less nervous about the interview.

Before the Interview

The following tips may seem basic, but you can be so focused on the interview itself that you lose sight of important fundamentals. In the hours before the interview, you must do the following:

1. Get a good night's sleep - Don't push your boundaries on the night before the interview. Allow yourself plenty of sleep so that you are alert on the day, and avoid caffeine or alcohol. Hopefully you've had plenty of time to prepare for the interview, and so you won't need to cram preparation the night before it.

2. Give yourself a lot of time - Even if the interview isn't until the afternoon, wake up early to prepare yourself. When you have at least two hours to get ready, you are going to be able to feel, look and arrive in the best way possible.

3. Groom – A smart appearance will show that you care about the role and that you're taking the interview seriously. It is advisable to do anything regarding personal hygiene that you would do for a formal occasion. Remember to also choose your clothes carefully. Don't overdo it – you don't need to wear something extremely formal – but you shouldn't arrive in a t-shirt and sandals. You need to make a good impression by making an effort; this will show that you truly want the job.

4. Eat something - There is nothing worse than a noisy stomach or fainting from low blood sugar during an interview. Make sure you eat well before the interview. Have a well-balanced meal two hours before the interview. Like the night before, absolutely avoid alcohol and preferably stay away from caffeine, too. Also remember to stay hydrated so that you can focus, but avoid drinking too much water – you don't want to be making constant trips to the toilet.

5. Give yourself travel time - Leave home early and arrive early in order to be absolutely sure that you are right on schedule. Also, arriving ahead of time gives you a chance to get ahead of the other candidates, to compose yourself before the interview, and to demonstrate that you are considerate of the committee. Arriving ahead of time will

leave a better impression than barely making it, and will obviously be much better than arriving late.

The Interview

We can't indicate what exact questions will be asked during any interview, but throughout your application process you have probably learned a lot about the individual company. This is going to prepare you well for the questions that they might reasonably ask. Remember too that all of the psychological profiling and team working skills have been assessed during the testing phase. Now, the committee is trying to get to know you. This is the time to use all of your preparation to give that one last push.

Consider that most committees will speak with you about:

- Relevant previous work experience and performance;

- Relevant community involvement;

- Your awareness of the role of a firefighter;

- Your ability to perform the role;

- Your ability to function as a member of a team.

This means you might get some questions that seem a bit blunt or direct, but just as in psychological profiling, they are not meant to antagonize. Often, understanding the underlying reason for any particular question will help you to answer more accurately. Your answers are also often compared to written answers you have given in the past. This is why honest has been the best policy from day one of the application process and onward.

Sample Interview Questions

Consider the following series of questions that are not uncommon during an interview for a firefighting position:

- Why do you want to be a firefighter?

- Have you ever done this sort of work?

- How often are you out of work due to sickness?

- Give an example of a fire or emergency situation in which you have been involved, and explain your actions during the situation.

- How would any of your previous colleagues describe you?

- Give an example of a time when you had to work as a member of a team. What did you learn?

- What do you know about this fire brigade/company/service?

- Will shift work present a problem to you?

- Have you ever worked in a team environment?

- Do you have difficulty with following a chain of command and/or taking orders when under emotional or physical strain?

- Have you ever had to pass a course in order to get a job before?

- How do you feel you handle physical fatigue?

- How do you feel you handle stress in your job?

- How do you prevent yourself from panicking during

moments of stress or danger?

These are all reasonable and common questions that a potential firefighter will be asked. Remember to not take the questions personally; the interview isn't designed to upset you or make personal attacks. The committee needs to decide if you are a good fit for the post and the team, which means that a range of questions will have to be asked.

After Your Interview

No matter what you feel about how the interview went, it's healthy to put it out of your mind. It usually takes weeks for a committee to make a decision, and so dwelling on how well or poorly you think you performed won't do you any favours. Regardless, the interview is over and second-guessing yourself won't change anything. Patience is key here, and you will receive a response to your hard work.

Most companies will communicate via mail or telephone. Generally, you are going to be told to report for your training session by an official postal communication. This will give you the documents and requirements that you must supply to register for the 12 or more weeks. All of the companies operate along a different model, so we cannot say if you have to plan for three months or more, but you will have to do the training according to the brigade's requirements.

What to take to the Interview

The most important things to bring to the interview are yourself, your confidence and a prepared state of mind. However, there are a few other things which may be useful on the day. These include:

- A pen and a notepad for taking notes. Taking this with you means that you can inquire about further details of the job, such as hours and compensation, and make a note of them. Additionally, you can prepare some questions prior to the interview and bring them with you to ask. You should plan to ask at least two or three questions because it demonstrates interest on your part, as well as that you are thinking carefully about the role rather than walking into it blindly. Asking a few further questions about the role should leave a good impression on the committee.

- A complete package of material about you. This includes the CV and all of the documents that have been essential to the process. It doesn't hurt to keep this with you.

In the final chapter, we will provide some interview questions, as well as some sample answers to inspire you.

Chapter Eight

Interview Questions and Answers

Here are some sample interview questions, as well as some exemplary answers. Use these to figure out how you would answer the interview questions. There are more sample questions listed in the previous chapter (p. 65) which you may wish to practice with once you've taken a look at these questions and answers.

Question 1

Tell me about a time when you have contributed to the effective working of a team.

How to structure your response:

• What was the size and purpose of the team?

• Who else was in the team?

• What was YOUR role in the team? (Explain your exact role)

• What did you personally do to help make the team effective?

• What was the result?

Strong response

To make your response strong you need to provide specific details of where you have worked with others effectively, and more importantly where YOU have contributed to the team. Try to think of an example where there was a problem within a team and where you volunteered to make the team work more efficiently. It is better to say that you identified there was problem within the team rather than that you were asked to do something by your manager or supervisor. Make your response concise and logical.

Weak response

Those candidates who fail to provide a specific example will provide weak answers. Do not fall into the trap of saying 'what you would do' if this type of situation arose.

Question 2

Tell me about a time when you helped someone who was distressed or in need of support.

How to structure your response:

• What was the situation?

• Why did you provide the help? (Whether you were approached or you volunteered – TIP: It is better to say you volunteered!)

• What did you do to support the individual?

• What specifically did you do or say?

• What was the result?

Strong response

Again, make sure you provide a specific example of where you have helped someone who was in distress or who needed your support. Try to provide an example where the outcome was a positive one as a result of your actions. If the situation was one that involved potentially dangerous surroundings (such as a car accident), did you consider the safety aspect and did you carry out a risk assessment of the scene?

Weak response

Candidates who provide a weak response will be generic in their answering. The outcome of the situation will generally not be a positive one.

Question 3

Describe a time when you have helped to support diversity in a team, school, college or organization.

How to structure your response:

• What was the situation?

• What prompted the situation?

• What were the diversity issues?

• What steps did you take to support others from diverse backgrounds?

• What specifically did you say or do?

• What was the result?

Strong response

This type of question is difficult to respond to, especially if you have little or no experience in this area. However, strong performing candidates will be able to provide clear details and examples of where they have supported diversity in a given situation. Their response will be specific in nature and it will clearly indicate to the panel that they are serious about this important subject.

Weak response

Weak responses are generic in nature and they fail to answer the question that is being asked. Many candidates are unable to provide a specific response to this type of question.

Question 4

Tell me about a time when you noticed a member of your team or group behaving in a manner which was inconsistent with the team's, group's, or organization's values.

How to structure your answer:

• What was the situation?

• How was the behavior inconsistent with the team's or organization's values?

• Why were the colleagues behaving in that way?

• What did you say or do when you noticed this behavior?

• What difficulties did you face?

• What was the result?

Strong response

Firefighters need to have the confidence and ability to challenge unacceptable behavior whilst at work. In order to understand what unacceptable behavior is, you first need to know what the values of the organization are. Candidates who provide a strong response will have a clear understanding of an organization's values and also how to tackle unacceptable behavior in the correct manner.

Weak response

Weak responses are generally where a candidate is unaware of the importance of an organization's values and how they impact on the needs of a team or group. They will not have the confidence to challenge inappropriate behavior and they will turn a blind eye whenever possible. Their response will lack structure and it will be generic in nature.

Question 5

Tell me about a time when you changed how you did something in response to feedback from someone else.

How to structure your response:

• What did you need to develop?

• What feedback did you receive and from whom?

• What steps did you take to improve yourself or someone else?

• What did you specifically say or do?

• What was the result?

Strong response

Firefighters receive feedback from their supervisory managers on a regular basis. In their quest to continually improve, the Fire Service will invest time, finances and resources into your development. Part of the learning process includes being able to accept feedback and also being able to improve as a result of it. Strong performing candidates will be able to provide a specific example of where they have taken feedback from an employer or otherwise, and used it to improve themselves.

Weak response

Those candidates who are unable to accept feedback from others and change as a result will generally provide a weak response to this type of question. They will fail to grasp the importance of feedback and in particular where it lies in relation to continuous improvement. Their response will be generic in nature and there will be no real substance or detail to their answer.

Question 6

Tell me about a time when you have taken it upon yourself to learn a new skill or develop an existing one.

How to structure your response:

• What skill did you learn or develop?

• What prompted this development?

• When did this learning or development occur or take place?

• How did you go about learning or developing this skill?

• What was the result?

• How has this skill helped you since then?

Strong response

Firefighters are required to learn new skills every week. They will attend ongoing training courses and they will also read up on new procedures and policies. In order to maintain a high level of professionalism, firefighters must be committed to continuous development. Try to think of an occasion when you have learnt a new skill, or where you have taken it upon yourself to develop your knowledge or experience in a particular subject. Follow the above structure format to create a strong response.

Weak response

Those candidates who have taken on any new development or learning will be unable to provide a strong response. They will provide a response where they were told to learn a new skill, rather than taking it upon themselves. There will be no structure to their learning or development and they will display a lack of motivation when learning.

Question 7

Tell me about a time when you had to follow clear instructions or rules in order to complete a task.

How to structure your response:

• What was the work you were doing?

• What were the rules or instructions that you had to follow?

• What did you do to complete the work as directed?

• What was the result?

• How did you feel about completing the task in this way?

Strong response

The Fire Service strives for excellence in everything it does. Therefore it is crucial that you provide a response that demonstrates you too can deliver excellence and maintain high standards. Try to think of a situation, either at work or otherwise, where you have achieved this. Make your response specific in nature. If you have had to follow specific instructions, rules or procedures then this is a good thing to tell the panel.

Weak response

Weak responses are generic in nature and usually focus on a candidate's own views on how a task should be achieved, rather than in line with a company's or organization's policies and procedures. The candidate will display a lack of motivation in relation to following clear instructions or rules.

Conclusion

In Conclusion

At this point, you have all of the details needed to pursue your career as a firefighter in Australia. We have covered the different brigades and companies operating in the different states and territories, the application process and selection process. We have also provided you with some materials to use when preparing for the written tests, as well as details on what to expect from the interview.

If you have followed all of our advice, filled the gaps in your application, made a fitness regime and taken the time to plan for the written tests and interview, you should be well on your way to becoming a strong candidate.

Bear in mind that firefighting is an extremely sought-after role, as well as incredibly demanding. For this reason, the selection process is rigorous and tough, and so you shouldn't be disheartened if you don't succeed in your early attempts. Persevere, follow the advice we have given you and remain confident in yourself as you go forward.

We wish you the best of luck and hope you quickly find the career of your dreams.

BONUS - As part of this guide, we would like you to try out our online training course now for FREE at:

https://www.how2become.com/courses/firefighter/

CHECK OUT OUR FIREFIGHTER RESOURCES

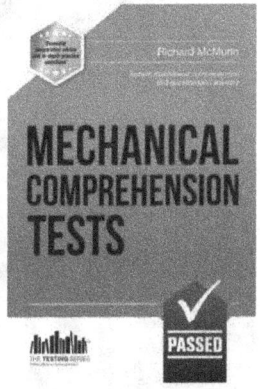

How2Become have created these other FANTASTIC guides to aid you through the Firefighter process.

FOR MORE INFORMATION ON OUR AUSTRALIAN FIREFIGHTER GUIDES, PLEASE CHECK OUT THE FOLLOWING:

WWW.HOW2BECOME.COM

FIREFIGHTER COURSE

- An online Firefighter course.

- Learn how to pass each stage of the Firefighter selection process.

- The course will cover the application form, the assessment centre and the interview.

WWW.HOW2BECOME.COM